Flash Revise Pocketbook

GCSE Chemistry
Science & Additional Science

Philip Allan Updates, an imprint of Hodder Education, an Hachette UK company, Market Place, Deddington, Oxfordshire OX15 0SE

Orders

Bookpoint Ltd, 130 Milton Park, Abingdon, Oxfordshire OX14 4SB
tel: 01235 827720   fax: 01235 400454   e-mail: uk.orders@bookpoint.co.uk

Lines are open 9.00 a.m.–5.00 p.m., Monday to Saturday, with a 24-hour message answering service. You can also order through our website: www.philipallan.co.uk

© Philip Allan Updates 2009
ISBN 978-0-340-99228-9

First published in 2006 as *Flashrevise Cards*

Impression number 5 4 3 2 1
Year 2014 2013 2012 2011 2010 2009

All rights reserved; no part of this publication may be reproduced, stored in a retrieval system, or transmitted, in any other form or by any means, electronic, mechanical, photocopying, recording or otherwise without either the prior written permission of Philip Allan Updates or a licence permitting restricted copying in the United Kingdom issued by the Copyright Licensing Agency Ltd, Saffron House, 6–10 Kirby Street, London EC1N 8TS.

Printed in Spain

Hachette UK's policy is to use papers that are natural, renewable and recyclable products and made from wood grown in sustainable forests. The logging and manufacturing processes are expected to conform to the environmental regulations of the country of origin.

P01538

Chemical reactions

1. Acid
2. Compound
3. Electrolysis
4. Limestone (calcium carbonate)
5. Sulfuric acid
6. Tests for gases and water
7. Tests for ions
8. Types of chemical reaction

Metals

9. Alloy
10. Blast furnace
11. Copper purification
12. Metal extraction
13. Ore
14. Transition elements

Atomic structure and the periodic table

15. Alkali metals
16. Atom
17. Carbon
18. Chlorine and its compounds
19. Displacement reaction between halogens
20. Electron configuration
21. Families of elements
22. Group 0 elements
23. Halogens
24. Isotopes
25. Periodic table

Rates of reaction

26. Collision theory

- **27** Enzymes
- **28** Rate of reaction

Bonding, structure and reacting quantities

- **29** Covalent bonding
- **30** Dot-and-cross diagram
- **31** Energetics
- **32** Ionic bonding
- **33** Relative formula mass

Oil and its products

- **34** Alkanes and alkenes
- **35** Combustion
- **36** Cracking
- **37** Crude oil
- **38** Oil industry
- **39** Polymer properties
- **40** Polymerisation

Earth cycles

- **41** Atmosphere
- **42** Carbon dioxide
- **43** Igneous rock
- **44** Metamorphic rock
- **45** Plate tectonics
- **46** Sedimentary rock

Equilibria and industrial processes

- **47** Ethanol
- **48** Haber process
- **49** Plant oils
- **50** Reversible reactions

GCSE Science & Additional Science: Chemistry
Chemical reactions

Acid

Q1 What is produced when an acid reacts with a carbonate?

Q2 What is produced when an acid reacts with reactive metals such as magnesium or zinc?

Q3 What is produced when an acid reacts with a base (or alkali)?

Q4 How does acid rain affect buildings?

substance that breaks down (dissociates) in water to form hydrogen ions (H⁺)

A1 A salt, carbon dioxide and water

A2 A salt and hydrogen, e.g.:
sulfuric acid + magnesium ⟶ magnesium sulfate + hydrogen

A3 A salt and water only, e.g.:
hydrochloric acid + sodium hydroxide ⟶ sodium chloride + water

A4 It corrodes metals and erodes statues and limestone building materials.

***exam* tip** You should know about hydrochloric acid (HCl, producing chlorides) and sulfuric acid (H_2SO_4, producing sulfates) in the reactions above. Acid rain contains dissolved carbon dioxide, sulfur dioxide and oxides of nitrogen. It is a weak acid.

ANSWERS

GCSE Science & Additional Science: Chemistry
Chemical reactions

2

Compound

Q1 How are the elements held together in compounds?

Q2 How many different types of element are there in H_2SO_4?

Q3 How many atoms are there in a molecule of H_2SO_4?

Q4 Do compounds and their constituent elements have similar properties?

ANSWERS

a substance made up of two or more elements chemically combined together

A1 Bonds — most are strong.

A2 Three: hydrogen, sulfur and oxygen

A3 Seven: two atoms of hydrogen, one of sulfur and four of oxygen

A4 No. For example, sodium chloride (table salt — a solid) has different properties from sodium (a very reactive metal) and chlorine (a poisonous green gas).

exam tip In a compound, the subscript numbers (e.g. the 2 in O_2) only refer to whatever is immediately in front of it. In $Mg(NO_3)_2$ there are two (NO_3) ions. There are nine atoms in total: the four within the bracket (one nitrogen and three oxygen) occur twice, giving eight atoms, plus the magnesium atom in front.

GCSE Science & Additional Science: Chemistry
Chemical reactions

Electrolysis

Q1 Why does an ionic substance conduct electricity?

Q2 What type of process occurs at each electrode?

Q3 If a mixture of ions is electrolysed, what determines which products are formed?

Q4 What is the half equation for the formation of chlorine by electrolysis?

ANSWERS

the decomposition of a substance by the passage of electricity

A1 When molten or in solution, the ions are free to move.

A2 Reduction at the cathode, oxidation at the anode

A3 The reactivity of the elements involved

A4 $2Cl^- \longrightarrow Cl_2 + 2e^-$

***exam* tip** The positive electrode is called the anode because it attracts negatively charged anions. The negative electrode is called the cathode because it attracts positively charged cations.

(3) **ANSWERS**

GCSE Science & Additional Science: Chemistry
Chemical reactions

Limestone (calcium carbonate)

Q1 Which processes use limestone as a raw material?

Q2 What impact does quarrying of limestone have on the environment?

Q3 Limestone is heated in a rotary kiln. What product is made and what is the product used for?

Q4 What is made when water is added to quicklime? What is its use?

sedimentary rock used in industry and agriculture

A1 Manufacture of cement and glass; also used as roadstone

A2 Visual impact (holes, pits, scars); noise (machinery); dust; destruction of natural habitats

A3 Quicklime (calcium oxide) — used in water treatment to neutralise acidity

A4 Slaked lime (calcium hydroxide) — used for neutralising excess acidity on land and for treating lakes affected by acid rain

exam tip You should know the equations for A3 and A4:

calcium carbonate ⟶ calcium oxide + carbon dioxide
$$CaCO_3 \longrightarrow CaO + CO_2$$
calcium oxide + water ⟶ calcium hydroxide
$$CaO + H_2O \longrightarrow Ca(OH)_2$$

 ANSWERS

GCSE Science & Additional Science: Chemistry
Chemical reactions

5

Sulfuric acid

Q1 Sulfur is burnt in air to produce sulfur dioxide. What is the symbol equation?

Q2 What catalyst is used to convert the sulfur dioxide to sulfur trioxide in the presence of more air?

Q3 Why is sulfur trioxide dissolved in oleum rather than being dissolved in water to make sulfuric acid?

Q4 What are the main uses of sulfuric acid?

ANSWERS

a strong acid with the formula H_2SO_4

A1 $S + O_2 \longrightarrow SO_2$

A2 Vanadium(V) oxide

A3 If water on its own is used, the reaction is too violent and produces a mist which is difficult to condense.

A4 Producing fertilisers, detergents, dyes, fibres, plastics and paints

exam tip The manufacture of sulfuric acid is known as the contact process. The equation for the catalysed reaction (see Question 2) is:
$$2SO_2 + O_2 \rightleftharpoons 2SO_3$$

GCSE Science & Additional Science: Chemistry
Chemical reactions

Tests for gases and water

Q1 Describe the test for hydrogen gas.

Q2 Describe the test for oxygen gas.

Q3 Describe the test for ammonia gas.

Q4 Describe the chemical tests for water.

laboratory procedures for identifying gases produced in reactions and water

A1 A lighted splint inserted into hydrogen gas will burn with a squeaky pop.

A2 A glowing splint inserted into oxygen gas will relight.

A3 Moist red litmus paper inserted into ammonia gas will turn from red to blue; ammonia is the only gas that will do this.

A4 • Water will turn anhydrous copper sulfate from white to blue.
   • Water will turn cobalt chloride paper from blue to pink.

***exam* tip** Don't get confused over the splint in Questions 1 and 2. A glowing splint will usually go out in hydrogen. A lighted splint in oxygen will burn brighter. Always give both the starting colour and the final colour when there is a colour change during a reaction.

GCSE Science & Additional Science: Chemistry
Chemical reactions

7

Tests for ions

Q1 What are the flame test colours for (a) copper, (b) potassium and (c) sodium?

Q2 What is the test for a carbonate?

Q3 How can aluminium ions be distinguished from calcium and magnesium ions?

Q4 How can chloride, bromide and iodide ions be distinguished?

an ion is a charged atom or group of atoms

A1 (a) Blue/green
   (b) Lilac
   (c) Bright orange/yellow

A2 Add acid, the gas produced turns limewater milky.

A3 They all produce a white precipitate with sodium hydroxide solution, but only the aluminium one will dissolve if excess hydroxide is added.

A4 Add silver nitrate solution. Silver chloride is white, silver bromide is pale yellow, silver iodide is darker yellow.

***exam* tip** Positively charged ions are called cations. Remember: 'cat-ions' are 'pussy-tive'. Negatively charged ions are called anions.

ANSWERS

GCSE Science & Additional Science: Chemistry
Chemical reactions

Types of chemical reaction

Q1 What is thermal decomposition?
Give an example.

Q2 What is neutralisation?
Give an example.

Q3 What happens when hydrogen is oxidised?
Give the equation.

Q4 What happens when a substance is reduced?
Give an example.

ANSWERS

there are four types: thermal decomposition, neutralisation, oxidation and reduction

A1 A substance is heated and it breaks down. For example:
calcium carbonate ⟶ calcium oxide + carbon dioxide

A2 An acid reacts with a base to produce a salt. For example:
sodium hydroxide + hydrochloric acid ⟶ sodium chloride + water

A3 Water is produced:
hydrogen + oxygen ⟶ water

A4 Oxygen is removed from the substance. For example:
zinc oxide + carbon monoxide ⟶ zinc + carbon dioxide

***exam* tip** Do not use the words 'exothermic' or 'endothermic' when giving examples of types of reaction; these simply give an indication of energy change.

 ANSWERS

GCSE Science & Additional Science: Chemistry
Metals

Alloy

Q1 Why are alloys harder than a pure metal alone?

Q2 Which soft metals are turned into alloys to make them harder for use?

Q3 What is a smart alloy?

Q4 Which other metals are added to iron to make stainless steel?

ANSWERS

a substance made by mixing a metal with one or more other metals or non-metals

A1 The different sized atoms prevent the layers of pure metal atoms from sliding easily.

A2 Aluminium, copper and gold

A3 An alloy that can return to its original shape after being deformed. Heat usually helps it regain its shape.

A4 Chromium and nickel

exam tip Most metals in everyday use are alloys. Bronze is an alloy of copper and tin; brass is an alloy of copper and zinc. 'Silver' coins are made from an alloy of copper and nickel.

GCSE Science & Additional Science: Chemistry
Metals

Blast furnace

Q1 What raw materials go into the blast furnace?

Q2 What is the symbol equation for the reaction?

Q3 Why is limestone added?

Q4 How is the iron treated to make steel?

industrial apparatus used to extract iron, the world's most widely used metal

A1 Iron ore (haematite), limestone and coke

A2 $Fe_2O_3 + 3CO \rightarrow 2Fe + 3CO_2$

A3 Heat breaks it down into calcium oxide which then reacts with acidic impurities in the ore (such as silicon dioxide), forming slag.

A4 Iron is brittle due to the high carbon content (about 4%) — oxygen is blown through to reduce the amount of carbon in the iron. Other metals are then added to confer special properties (e.g. chromium for making stainless steel).

exam tip Be prepared to label a diagram of the blast furnace (remember that the slag floats on the molten iron). You must know the main symbol equation.

GCSE Science & Additional Science: Chemistry
Metals

Copper purification

Q1 What is the anode made of?

Q2 Where does the pure copper deposit? What is the electrolyte?

Q3 Write out the half equations for the reactions at each electrode?

Q4 What would you expect to see during this reaction?

electrolysis of impure copper to extract the pure metal

A1 The anode (positive electrode) is made of impure copper.

A2 Pure copper collects on the cathode; the electrolyte is copper sulfate solution.

A3 • Cathode (negative electrode): $Cu^{2+}(aq) + 2e^- \longrightarrow Cu(s)$
 • Anode (positive electrode): $Cu(s) \longrightarrow Cu^{2+}(aq) + 2e^-$

A4 The cathode would become thicker and the anode would decrease in size. The colour of the electrolyte would not alter.

exam tip Copper used for electrical wiring must be extremely pure (>99.995%). Impurities would increase the resistance of the copper too much. During the electrolysis, impurities collect under the anode. They include valuable metals such as gold, silver and platinum.

GCSE Science & Additional Science: Chemistry
Metals

12

Metal extraction

Q1 What technique is required to extract a metal that is high in the reactivity series from its ore?

Q2 What technique is required to extract a metal that is in the middle of the reactivity series from its ore?

Q3 What technique is required to extract a metal that is at the bottom of the reactivity series from its ore?

Q4 What type of process is the extraction of a metal from its ore?

production of pure metal from its ore, using a range of techniques

- **A1** Electrolysis — usually of the fused (molten) ore, e.g. aluminium
- **A2** Reducing — usually with carbon or carbon monoxide, e.g. the preparation of iron from iron ore
- **A3** Heating — e.g. heating mercury oxide produces mercury and oxygen
- **A4** Reduction — the metal gains electrons during the extraction process

***exam* tip** You must be able to use a metal's position in the reactivity series to suggest a method of extraction.

GCSE Science & Additional Science: Chemistry
Metals

Ore

Q1 Ores containing which metallic element are most common in the Earth's crust?

Q2 What determines the technique used to extract metals from their ores?

Q3 What needs to happen before the metal can be released from its ore?

Q4 Can metal extraction go on for ever?

> rock containing minerals (metal compounds), or metals, from which it is economical to extract the metal

- **A1** Al (aluminium), which makes up 8% of the Earth's crust.
- **A2** The position of the metal in the reactivity series. The higher the metal in the series, the more strongly it will hold onto its oxide, so more forceful methods will be needed to extract it.
- **A3** The ore needs to be purified; other substances must be removed so that concentrated ore remains.
- **A4** No. Some metal ores are in limited supply, and all of the Earth's mineral resources are finite.

exam tip Recycling can help to prolong the lifetime of a metal resource. Recycling metals uses up less energy than extracting and purifying the metal from its ore.

GCSE Science & Additional Science: Chemistry
Metals

14

Transition elements

Q1 How do the melting points of transition elements differ from those of other metals?

Q2 How does the density of transition elements compare with that of other metals?

Q3 In what way are the colours of transition element compounds different from those of other metallic elements?

Q4 What are (a) iron, (b) nickel and (c) copper used for?

ANSWERS

the block of elements inserted between groups 2 and 3 in the periodic table

A1 They are high — although there are exceptions, such as mercury.

A2 Transition elements are more dense than other metals.

A3 Compounds of transition elements are coloured, whereas compounds of other main block metals are white. Predominant colours are: iron — orange or pale green; nickel — green; and copper — blue.

A4 (a) Iron is used as a structural material and as a catalyst in the Haber process for making ammonia. (b) Nickel is used in coins and as a catalyst in the manufacture of margarine. (c) Copper is used in wiring and for making water pipes.

exam tip Transition metals can have more than one valency. The written name tells you what it is, for example copper(I) oxide (Cu_2O) and copper(II) oxide (CuO). You only need to be concerned with iron, nickel and copper at GCSE.

GCSE Science & Additional Science: Chemistry
Atomic structure and the periodic table

15

Alkali metals

Q1 What are the characteristic physical properties of group 1?

Q2 What is produced when the alkali metals react with water?

Q3 Why do the alkali metals become more reactive as you descend the group?

Q4 What safety precautions should be taken when handling the alkali metals?

ANSWERS

the elements in group 1: lithium, sodium, potassium, rubidium, caesium and francium

A1
- All the elements in group 1 are moderately soft metals (increasing in softness down the group).
- Their densities increase on descending the group (lithium, sodium and potassium float on water; the others sink).
- Their melting points decrease on descending the group.

A2 The metal hydroxide and hydrogen gas

A3 The outermost electron is furthest from the nucleus and thus shielded by the other electrons, so is more easily lost.

A4 Wear safety glasses, do reactions behind a safety screen and only use small amounts.

exam tip Metals always form positive ions when they react as they lose electrons from their outermost shell to attain a full outer shell of electrons.

GCSE Science & Additional Science: Chemistry
Atomic structure and the periodic table

16

Atom

Q1 What are atoms made of and how are they arranged?

Q2 What are the relative charges of (a) electrons, (b) protons and (c) neutrons?

Q3 What are the relative masses of (a) electrons, (b) protons and (c) neutrons?

Q4 What is meant by the atomic number of an element?

ANSWERS

the smallest part of an element that retains the properties of that element

A1 Atoms consist of a central nucleus made up of protons and neutrons, surrounded by orbiting electrons.

A2 (a) Electrons — negative
(b) Protons — positive
(c) Neutrons — neutral

A3 (a) Electron — negligible
(b) Proton — relative mass of 1
(c) Neutron — relative mass of 1

A4 It is the number of protons in the nucleus of the element.

***exam* tip** Different elements have different numbers of protons in their nucleus. If the number of protons changes, so does the element.

GCSE Science & Additional Science: Chemistry
Atomic structure and the periodic table

17

Carbon

Q1 Which of the three allotropes of carbon are giant structures?

Q2 Why does graphite conduct electricity, but diamond does not?

Q3 Why is graphite soft but diamond is the hardest natural material known?

Q4 Why are the melting points of diamond and graphite so high?

can exist in a number of different forms (allotropes) — diamond, graphite and buckminsterfullerene

A1 Diamond and graphite (buckminsterfullerene is a molecule).

A2 Carbon atoms in graphite are bonded to only three others, so there is an unbonded electron available to help conduct electricity. In diamond, all four outer electrons are bonded to other carbon atoms.

A3 In graphite, the atoms are arranged in layers with weak forces between them. In diamond, the atoms are held in a rigid three-dimensional structure.

A4 The carbon–carbon bonds are very strong and so need a lot of energy to break them.

***exam* tip** You will not need to be able to draw the structures of diamond, graphite or buckminsterfullerene — however, you will need to be able to tell them apart from diagrams and pictures.

GCSE Science & Additional Science: Chemistry
Atomic structure and the periodic table

Chlorine and its compounds

Q1 Why is solid sodium chloride used on roads in winter?

Q2 What is manufactured from sodium chloride?

Q3 What is chlorine used for?

Q4 What does CFCs stand for? Why should we worry about them?

ANSWERS

a reactive element, which reacts readily with metals; its compounds are widely used

A1 It lowers the freezing point of water and so prevents ice from forming.

A2 Sodium hydroxide and chlorine (by the electrolysis of sodium chloride)

A3 Sterilising water supplies and making bleach (NaClO) are the two most important uses; it is also used for the manufacture of plastics, pesticides, weed killers, solvents and hydrochloric acid.

A4 Chlorofluorocarbons. They were used in the 1960s and 70s as refrigerants, in aerosols, and for expanding polystyrene. They last for many centuries in the atmosphere but can be broken down by ultraviolet light into reactive species that destroy the ozone layer.

***exam* tip** Most chlorine is made by the electrolysis of brine from salt mining.

GCSE Science & Additional Science: Chemistry
Atomic structure and the periodic table

19

Displacement reaction between halogens

Q1 In which of the following will a reaction occur, and why?
(a) bromine(aq) + sodium chloride(aq)
(b) bromine(aq) + sodium iodide(aq)

Q2 What colour changes would be seen?

Q3 What safety precautions should be taken if chlorine is prepared in the lab?

Q4 What safety precautions should be taken if iodine is used in the lab?

displacement of a halogen by a more reactive halogen from a solution of its salt

A1 Reaction (b), since bromine is more reactive than iodine:
bromine(aq) + sodium iodide(aq) ⟶
iodine(aq) + sodium bromide(aq)

A2 The orange/red bromine solution would turn purple as iodine is produced; sodium bromide is colourless in water.

A3 Wear safety glasses and perform the reaction in a fume cupboard, since chlorine is a choking toxic gas.

A4 Iodine is a toxic solid that can cause burns, so safety glasses should be worn and a spatula used. Sodium thiosulfate solution can be used to neutralise it.

***exam* tip** In displacement reactions, an immiscible liquid is usually added so that the released halogen dissolves in it, to better observe the colour changes.

GCSE Science & Additional Science: Chemistry
Atomic structure and the periodic table

20

Electron configuration

Q1 How many electrons can each of the first three shells hold?

Q2 Which shell determines chemical properties?

Q3 What is the shorthand notation (e.g. Li: 2, 1) for (a) potassium, (b) chlorine and (c) aluminium?

Q4 Draw electron shell diagrams for (a) beryllium, (b) fluorine and (c) carbon.

ANSWERS

the arrangement of electrons in shells around a nucleus

A1 The first (inner) shell holds a maximum of two electrons, the second and third shells a maximum of eight electrons each.

A2 The inner shells are full and stable; it is the outer, partially full, shell that determines the chemical properties.

A3 (a) K: 2, 8, 8, 1; (b) Cl: 2, 8, 7; (c) Al: 2, 8, 3

A4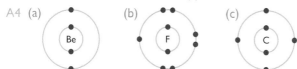

exam tip You must be able to work out the electron configuration for the first 20 elements. However, you may be asked how many electrons there are in the outer shell of *any* main group element — this is the same as the group number.

GCSE Science & Additional Science: Chemistry
Atomic structure and the periodic table

Families of elements

Q1 What is a group and what position does it occupy in the periodic table?

Q2 What are the common names for group 1 and group 7?

Q3 Prior to Mendeleev, the periodic table was arranged according to atomic mass. What can you say about the positions of potassium and argon?

Q4 How did Mendeleev organise the periodic table?

the grouping of elements depending on their chemical and physical properties

A1 A group is a family of elements that runs vertically down the periodic table. All elements in the group have similar properties, as they all have the same number of outer electrons.

A2 Group 1 — alkali metals; group 7 — halogens

A3 They would be exchanged — potassium would be in group 8 and argon would be in group 1. This would not match the properties of the other members of these groups.

A4 He arranged the elements in order of atomic number.

exam tip The last group in the periodic table, the noble gases, is given the group number 0. It is, however, the eighth group in the table, which is useful to remember when working out how many electrons there are in the outer shell of each of the noble gases (i.e. eight) — apart from helium, which has two.

GCSE Science & Additional Science: Chemistry
Atomic structure and the periodic table

22

Group 0 elements

Q1 Why are the group 0 elements unreactive?

Q2 How does the density of the gases change as the group is descended?

Q3 What is unusual about the group 0 gas particles compared with other gaseous elements in the periodic table?

Q4 What are the group 0 elements used for?

ANSWERS

commonly known as the noble gases; they are very unreactive

A1 They have a full outer shell of electrons and so do not need to lose or gain electrons in order to complete the outer shell.

A2 The gases become more dense on descending the group.

A3 The group 0 gases are monatomic; all other gaseous elements are diatomic.

A4
- Helium in balloons
- Neon in advertising signs
- Argon in light bulbs
- Krypton in lasers

***exam* tip** When elements bond, they do so in order to gain the same electron arrangement as the group 0 elements, since this is a stable arrangement.

GCSE Science & Additional Science: Chemistry
Atomic structure and the periodic table

Halogens

Q1 Give the colour and state of (a) fluorine, (b) chlorine, (c) bromine and (d) iodine.

Q2 How do the melting and boiling points change on descending the group?

Q3 How does the reactivity of halogens with metal change on descending the group?

Q4 Explain your answer to Q3.

the elements in group 7 — they are all non-metals and form ions with a charge of –1

A1 (a) Pale yellow gas; (b) green gas; (c) orange/red liquid; (d) shiny grey solid

A2 Both melting and boiling points increase on descending the group.

A3 Halogens become less reactive as the group descends.

A4 Halogens only need one more electron — accepted from the metal — to complete their outer shell. With fluorine, the added electron is held closer and more strongly to the nucleus, whereas for the other elements it is further from the nucleus and therefore held less strongly, so the reactivity with metals is lower.

exam tip If asked about the properties of astatine, use the trends observed from the other halogens to guide your answer. From these trends, astatine will be a black solid with a high melting point and will not be very reactive.

GCSE Science & Additional Science: Chemistry
Atomic structure and the periodic table

Isotopes

Q1 Using a periodic table, give the mass numbers of (a) argon, (b) potassium and (c) aluminium.

Q2 How are the numbers of protons and electrons related?

Q3 Why do isotopes of an element differ in mass?

Q4 How many protons, neutrons and electrons are there in an atom of (a) $^{23}_{11}Na$, (b) $^{31}_{15}P$ and (c) $^{80}_{35}Br$?

ANSWERS

elements with the same number of protons, but different numbers of neutrons

A1 (a) Argon = 40; (b) potassium = 39; (c) aluminium = 27

A2
- In a neutral atom there are equal numbers of protons and electrons.
- In a cation there are fewer electrons than protons.
- In an anion there are more electrons than protons.

A3 The atoms vary in the number of neutrons — each neutron has a relative mass of 1.

A4 (a) Na: 11 protons, 12 neutrons, 11 electrons
(b) P: 15 protons, 16 neutrons, 15 electrons
(c) Br: 35 protons, 45 neutrons, 35 electrons

exam tip On a periodic table, the largest number by an element is the mass number — the sum of protons and neutrons. The lowest number is the atomic number (the number of protons).

GCSE Science & Additional Science: Chemistry
Atomic structure and the periodic table

Periodic table

Q1 (a) What are vertical columns of elements called in the periodic table? (b) What are horizontal rows of elements called?

Q2 How is the number of outer electrons of an element related to its position in the periodic table?

Q3 What can you say about the properties of elements in any particular group?

Q4 Give an explanation for your answer to Q3.

an arrangement of all known elements in order of increasing atomic number

A1 (a) Vertical columns are called groups.
   (b) Horizontal rows are called periods.
A2 The number of the group the element is in corresponds to the number of outer electrons for that element.
A3 All the elements in a group have similar chemical properties.
A4 They have the same number of outer electrons to react and it is the outer electrons that determine the reactivity of an element.

***exam* tip** Metallic elements increase in reactivity on descending the group. Non-metallic elements increase in reactivity on ascending the group.

GCSE Science & Additional Science: Chemistry
Rates of reaction

Collision theory

Q1 What must happen to particles in a reaction in order that they may react?

Q2 Does a reaction always occur between colliding particles?

Q3 How does increasing the concentration of solutions affect the rate of reaction?

Q4 How does increasing the temperature affect the rate of reaction?

describes the conditions needed for reactions to take place between particles

A1 They must collide in the correct orientation.

A2 No, they may not have enough energy for the reaction to be successful.

A3 There will be more collisions between particles and therefore more successful collisions.

A4 Increasing the temperature increases the energy of the particles, leading to more successful collisions.

***exam* tip** Remember that the finer the particles of reacting solids, the larger the surface area they have for reactions to take place and the faster the reaction will be.

GCSE Science & Additional Science: Chemistry
Rates of reaction

27

Enzymes

Q1 How does enzyme activity vary with temperature?

Q2 How do high temperatures affect the activity of an enzyme?

Q3 How does pH affect enzyme activity?

Q4 What are enzymes used for?

ANSWERS

biological catalysts produced by organisms; they are specific in the reactions they catalyse

A1 It increases up to an optimum temperature and then decreases again.

A2 The enzyme becomes denatured — its active site will no longer accept molecules in the correct orientation for reactions to take place.

A3 Enzymes are very sensitive to pH — any changes can stop the enzyme from working properly.

A4 • Brewing
- Biological detergents
- Softening the centres of some chocolates (using invertase)

exam tip You should know the equation for fermentation in the presence of yeast:

sugar ⟶ ethanol + carbon dioxide
$C_6H_{12}O_6$ ⟶ $2C_2H_5OH$ + $2CO_2$

 ANSWERS

GCSE Science & Additional Science: Chemistry
Rates of reaction

28

Rate of reaction

Q1 Give some examples of different reaction rates.

Q2 How can a reaction rate be followed?

Q3 What factors affect the rate of reaction?

Q4 How does a catalyst affect the rate of a reaction?

ANSWERS

the speed at which a reaction takes place

A1
- Fast: explosions; combustion
- Medium: reactions of metals with acids; decomposition of hydrogen peroxide
- Slow: rusting
- Very slow: chemical weathering of rocks

A2 By seeing how quickly the reactants are used up, or how quickly the products are formed

A3 The concentration of reactants in solution; the temperature; the size of any solid pieces; the pressure of gases present

A4 Catalysts speed up reactions by lowering the amount of energy needed to start the reaction (the activation energy).

***exam* tip** Catalysts speed up reactions but do not affect the overall yield.

GCSE Science & Additional Science: Chemistry
Bonding, structure and reacting quantities

29

Covalent bonding

Q1 Are the forces between molecules strong or weak?

Q2 What physical properties do most covalently bonded substances share?

Q3 What causes the molecules to have these properties?

Q4 Give some examples of giant molecular structures.

ANSWERS

bonding between atoms through the sharing of electrons

A1 There are only weak forces of attraction between molecules.

A2 They are usually gases, liquids or, if solid, have a low melting point.

A3 Since the forces between the molecules are weak, not much energy is needed in order to disrupt the structure.

A4 DNA, polymers, diamond and graphite are examples of giant molecular structures.

***exam* tip** Remember that in ionic compounds the forces between ions allow large extended structures to be built up, whereas for molecular substances, with weak forces between molecules, there is a finite size to the molecule.

GCSE Science & Additional Science: Chemistry
Bonding, structure and reacting quantities

30

Dot-and-cross diagram

Q1 How are the outer electrons arranged in hydrogen (H_2)?

Q2 How are the outer electrons arranged in chlorine (Cl_2)?

Q3 How are the outer electrons arranged in water (H_2O)?

Q4 How are the outer electrons arranged in oxygen (O_2)?

illustration of covalent bonding within a molecule

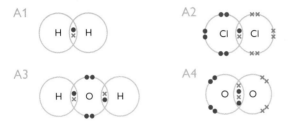

exam tip Make sure that there are a total of eight electrons around each atom apart from hydrogen, which should only have a total of two in its outer shell after bonding.

GCSE Science & Additional Science: Chemistry
Bonding, structure and reacting quantities

Energetics

Q1 In an energy level diagram for an exothermic reaction, do the products have more or less energy than the reactants?

Q2 Are all bond strengths the same?

Q3 What energy changes are associated with making and breaking bonds?

Q4 During a reaction, how does the making and breaking of bonds affect the overall energy of the reaction?

ANSWERS

energy changes, usually in the form of heat, that accompany chemical reactions

A1 Less energy — as some has been given out during the reaction

A2 No. Bond strengths differ according to the two elements that are bonded together.

A3 • Energy is needed to break a bond — endothermic process
   • Energy is given out when bonds are made — exothermic process

A4 If more energy is given out when the product bonds are made than was taken in to break the reactant bonds, then the overall reaction is exothermic; and vice versa.

***exam* tip** Generally, the further apart the elements are in the periodic table, the stronger the bond between them.

ANSWERS

GCSE Science & Additional Science: Chemistry
Bonding, structure and reacting quantities

32

Ionic bonding

Q1 Show what happens when sodium loses electrons.

Q2 Show what happens when chlorine gains electrons.

Q3 How many electrons are transferred from the metal to the non-metal?

Q4 What holds the charged particles together?

ANSWERS

bonding between a metal and a non-metal — electrons are transferred from the metal to the non-metal

A1 The metal becomes positively charged since it now has more protons than electrons:

$$Na \longrightarrow Na^+ + e^-$$

A2 The non-metal becomes negatively charged since it now has more electrons than protons:

$$Cl + e^- \longrightarrow Cl^-$$

A3 As many as are needed to fill the outer shell of electrons

A4 Strong electrostatic forces hold the particles together in a large three-dimensional network (lattice).

exam tip You only have to know about combinations of the following elements: Li, Na or Mg with F, Cl or O.

GCSE Science & Additional Science: Chemistry
Bonding, structure and reacting quantities

Relative formula mass

Q1 Calculate the relative formula masses of (a) H_2SO_4 and (b) $CaCO_3$. (r.a.m. H = 1, C = 12, O = 16, S = 32, Ca = 40)

Q2 Calculate the percentage by mass of nitrogen in ammonium nitrate (NH_4NO_3). (r.a.m. H = 1, N = 14, O = 16)

Q3 What mass of calcium oxide can be obtained from heating 100 g of calcium carbonate?

Q4 What mass of calcium sulfate ($CaSO_4$) is produced by adding sulfuric acid (H_2SO_4) to 16.4 g of calcium nitrate ($Ca(NO_3)_2$)?

mass of a molecule compared with $\frac{1}{12}$ of the mass of an atom of carbon-12

A1 (a) 98
   (b) 100

A2 mass of nitrogen = 28; mass of ammonium nitrate = 80
   percentage by mass = $(28/80) \times 100 = 35\%$

A3 Equation: $\quad CaCO_3 \longrightarrow CaO + CO_2$
   Formula masses: $\quad\;\;\, 100 \qquad\;\; 56 \quad\; 44$
   Hence mass of CaO = 56 g

A4 Equation: $\quad Ca(NO_3)_2 + H_2SO_4 \longrightarrow CaSO_4 + 2HNO_3$
   Formula masses: $\;\; 164 \qquad\;\; 98 \qquad\quad\; 136 \quad\;\; 2(63)$
   Use ratios to find the mass of product: $(16.4/164) \times 136 = 13.6\,g$

exam tip You can check that you have balanced an equation correctly if the total formula masses of the reactants equals the total formula masses of the products.

ANSWERS

GCSE Science & Additional Science: Chemistry
Oil and its products

34

Alkanes and alkenes

Q1 What is the main structural difference between an alkane and an alkene?

Q2 What are the names and formulae of the first three alkanes and the first two alkenes?

Q3 What are the general formulae for alkanes and alkenes?

Q4 What is the test used to distinguish between an alkane and an alkene?

ANSWERS

two different families of hydrocarbon

A1
- Alkanes only contain single bonds between the carbon atoms.
- Alkenes contain one or more double bond between carbon atoms (C=C).

A2
- Alkanes: methane (CH_4), ethane (C_2H_6) and propane (C_3H_8)
- Alkenes: ethene (C_2H_4) and propene (C_3H_6)

A3
- Alkanes: $C_nH_{(2n+2)}$
- Alkenes: C_nH_{2n}

A4 Shake the hydrocarbon with bromine water (orange); alkanes have no effect on the colour whereas alkenes decolorise the bromine water.

***exam* tip** Alkanes are known as 'saturated' compounds and alkenes are known as 'unsaturated'. Beware — this has nothing to do with water content!

GCSE Science & Additional Science: Chemistry
Oil and its products

Combustion

Q1 What are the products of the complete combustion of hydrocarbons?

Q2 What is the symbol equation for the complete combustion of ethane (C_2H_6)?

Q3 Carbon monoxide is formed by incomplete combustion of hydrocarbons. Why is this a problem?

Q4 Most fuels contain some sulfur. Why is this a problem?

process whereby a substance combines with oxygen to produce heat and light

A1 Carbon dioxide and water only

A2 $2C_2H_6 + 7O_2 \longrightarrow 4CO_2 + 6H_2O$

A3 Carbon monoxide is odourless, colourless and toxic (it binds irreversibly to haemoglobin in the blood).

A4 On combustion sulfur dioxide is formed, which can contribute to acid rain.

***exam* tip** During combustion, water is always formed as one of the products — it is the oxide of carbon that varies.

GCSE Science & Additional Science: Chemistry
Oil and its products

36

Cracking

Q1 What are the conditions needed for cracking to occur?

Q2 What are the products of cracking?

Q3 Why are long alkanes cracked?

Q4 What type of reaction do alkenes undergo?

breaking down long-chain hydrocarbons into shorter-chain alkanes and alkenes

A1 Heat or a hot catalyst

A2 Shorter alkanes and alkenes are produced; depending on conditions, hydrogen may also be produced.

A3 Commercially they are not very useful; the shorter alkenes produced can be used as the basis for making many other chemicals or polymers.

A4 Addition reactions — two reactants produce only one product.

***exam* tip** When writing equations, remember that an alkane will always give a shorter alkane and an alkene.

GCSE Science & Additional Science: Chemistry
Oil and its products

37

Crude oil

Q1 Under what conditions is oil formed?

Q2 What is a hydrocarbon?

Q3 How is crude oil split up into its separate components?

Q4 What are the main uses of the fractions of crude oil?

a natural resource consisting mostly of a mixture of hydrocarbons; it is finite and non-renewable

A1 Organic material is buried in an oxygen-free environment for a long period of time, under pressure, at a temperature of around 90–120°C.

A2 An organic molecule made up of carbon and hydrogen atoms only

A3 Fractional distillation

A4 • Refinery gases for bottled fuel
- Gasoline for cars
- Kerosine for jet aircraft
- Diesel for cars and lorries
- Fuel oil for heating systems

***exam* tip** Remember that hydrocarbons with few carbon atoms come off near the top of the column — they have low boiling points; hydrocarbons with high numbers of carbon atoms come off near the bottom — they have high boiling points.

37 ANSWERS

GCSE Science & Additional Science: Chemistry
Oil and its products

38

Oil industry

Q1 What is the impact on the environment of extracting crude oil?

Q2 Why is it difficult to recycle polymers?

Q3 What other environmental problems can polymers cause?

Q4 Is oil a renewable resource?

important source of fuels and raw materials for other commodities (e.g. polymers and pharmaceuticals)

A1 Oil spills and discharges can pollute rivers and oceans and harm wildlife.

A2 There are many different types of polymer and it is difficult to separate them when they are being recycled. A single object can be made out of many different polymers.

A3 Most polymers are not biodegradable; they can potentially last in the environment for many hundreds of years.

A4 Oil is a non-renewable resource; we are using it up much more quickly than natural processes can make it.

exam **tip** Don't forget that the oil industry requires many different types of scientist to work for it. They are needed for the detection, extraction and purification of oil, as well as for the development of its uses.

GCSE Science & Additional Science: Chemistry
Oil and its products

39

Polymer properties

Q1 What is the polymer made from ethene used for?

Q2 What is the polymer made from propene used for?

Q3 What problems can polymers cause to the environment?

Q4 What new uses have been found for polymers?

ANSWERS

mechanical characteristics dictated by the monomer and the nature of its bonding

A1 Poly(ethene) is used for plastic bags and bottles as it is a flexible polymer.

A2 Poly(propene) is used for crates and rope — in bulk form it is fairly rigid, but it can be spun into fibres for making rope and carpet.

A3 Most polymers are not biodegradable so they cannot be broken down by microorganisms; this can lead to pollution.

A4 New packaging materials, waterproof coatings for fabrics, dental polymers, wound dressings, shape memory polymers and hydrogels

exam **tip** All these polymers are addition polymers — the monomers contained double bonds. Apart from the polymer, no other product is produced when the polymer is made.

GCSE Science & Additional Science: Chemistry
Oil and its products

Polymerisation

Q1 What is the difference between a monomer and a polymer?

Q2 How do you identify the monomer that makes up a polymer?

Q3 Write the symbol equation for the polymerisation of ethene.

Q4 How are addition polymers named?

alkenes reacting with themselves to produce long-chain polymer molecules

A1 A monomer is the building block (an alkene) from which the polymer is made.

A2 Find the part of the polymer that keeps repeating itself — this is the alkene from which the polymer is made.

A3 $n(C_2H_4) \longrightarrow -(CH_2-CH_2)_n$ (where n stands for a large number)

A4 They are named after the monomer (i.e. as poly(alkene)), even though the polymer does not contain a double bond.

***exam* tip** Always use the term 'polymer', not 'plastic'. The word plastic means something quite different.

GCSE Science & Additional Science: Chemistry
Earth cycles

Atmosphere

Q1 What is the composition of the major gases in the atmosphere?

Q2 How can you show that air is approximately one-fifth oxygen?

Q3 What was the composition of the Earth's first atmosphere?

Q4 What caused the build up of oxygen in the atmosphere?

layer of mixed gases that surrounds the Earth

A1
- 78% nitrogen (about four-fifths)
- 21% oxygen (about one-fifth)
- 0.9% argon
- 0.03% carbon dioxide and smaller amounts of other gases

A2 Pass a fixed volume of air over heated copper using gas syringes — the volume of gas in the syringes will decrease by one fifth.

A3 It was mostly carbon dioxide and water vapour, with smaller amounts of methane and ammonia.

A4 Photosynthesising organisms on land and in the sea.

***exam* tip** The early atmosphere was formed by intense volcanic activity, which released the gases.

GCSE Science & Additional Science: Chemistry
Earth cycles

42

Carbon dioxide

Q1 What caused the reduction of carbon dioxide in the Earth's early atmosphere?

Q2 Why is carbon dioxide starting to build up again in the atmosphere?

Q3 How do the oceans act as a reservoir for carbon dioxide?

Q4 Carbon dioxide in the atmosphere is a greenhouse gas. Explain what this means.

colourless, odourless gas produced when carbon-containing compounds are burned in air

A1
- It was removed by photosynthesis.
- It dissolved in the ocean.
- It eventually helped form fossil fuels once it had been removed by plants.

A2 The burning of fossil fuels since the industrial revolution has increased the amount of carbon dioxide being emitted into the atmosphere.

A3 Carbon dioxide dissolves in the oceans, forming a very weak solution of carbonic acid (H_2CO_3), and is thus removed from the atmosphere.

A4 Carbon dioxide can trap infrared energy radiated from the Earth, causing the atmosphere to warm up — this is known as global warming.

exam tip Don't confuse global warming due to greenhouse gases with the increase of ultraviolet light reaching the Earth due to depletion of the ozone layer.

GCSE Science & Additional Science: Chemistry
Earth cycles

43

Igneous rock

Q1 What is the general appearance of an igneous rock?

Q2 What causes the formation of granite?

Q3 What causes the formation of basalt?

Q4 What is meant by intrusive and extrusive when applied to igneous rocks?

ANSWERS

rock formed by the cooling of molten magma

A1 It is chunky, massive rock formed from interlocking crystals.

A2 Granite is formed when magma solidifies under the ground. It cools slowly, so large interlocking crystals are formed.

A3 Basalt is formed when magma cools on the surface of the Earth. It cools quickly, so there is not enough time for large crystals to form; small interlocking crystals only are observed.

A4 • Granite is intrusive as it was formed within the Earth.
  • Basalt is extrusive as it was formed on the surface of the Earth.

***exam* tip** The word igneous literally means 'born of fire'.

 ANSWERS

GCSE Science & Additional Science: Chemistry
Earth cycles

Metamorphic rock

Q1 What is the general appearance of metamorphic rock?

Q2 What type of metamorphism occurs when a rock is next to hot magma?

Q3 What is regional metamorphism?

Q4 What is marble?

ANSWERS

rock formed whenever heat and pressure are applied to other rocks over a long period

A1 It usually contains both layers and crystals.

A2 Contact metamorphism — the rocks nearest the heat source will change the most.

A3 This metamorphism is caused when both heat and pressure are found over large areas due to mountain building processes or plate tectonic movements.

A4 Marble is a metamorphic rock formed by the metamorphism of limestone or chalk; some marble may still contain traces of fossils.

***exam* tip** Metamorphism may be 'low grade' if there has not been much heat and pressure (e.g. forming slate from shale) or 'high grade' if there has been a lot of heat and pressure (e.g. schist and gneiss from shale).

GCSE Science & Additional Science: Chemistry
Earth cycles

Plate tectonics

Q1 In a layered series of rocks, which is the oldest?

Q2 When one rock type cuts across another, which is the younger?

Q3 What other factor can help to date sedimentary rocks?

Q4 What causes rocks to fault and fold?

ANSWERS

theory that the Earth's crust is made up of a number of moving plates

A1 The oldest rock is always found on the bottom — if the strata are horizontal — with progressively younger rocks on top of it.

A2 The one that cuts across is the younger.

A3 Fossil evidence can be used to date rocks — by knowing when an animal or plant lived in the past.

A4 The movement of major tectonic plates bumping into each other, pulling apart or rubbing past each other.

***exam* tip** Tectonic plates move at the rate of a growing fingernail — about 5 cm a year.

GCSE Science & Additional Science: Chemistry
Earth cycles

Sedimentary rock

Q1 What is the general appearance of a sedimentary rock?

Q2 What is the process by which the fine grains of rock are 'glued' together known as?

Q3 What is shale?

Q4 Limestone ($CaCO_3$) is a sedimentary rock. How is it formed?

rock formed when weathered rock settles as sediment and is then subjected to high pressure

A1 It is usually formed in layers; there are no interlocking crystals visible.

A2 Cementation or lithification

A3 Shale is fine-grained rock formed from the compression of mud and silt laid down in river estuaries.

A4 Limestone is formed from the skeletons and shells of sea creatures compressed and cemented over millions of years.

exam tip Remember that although sedimentary rocks are formed on the ocean floor, uplift or sea-level changes will have brought them above sea level to where we can see them now.

GCSE Science & Additional Science: Chemistry
Equilibria and industrial processes

Ethanol

Q1 How is sugar turned into ethanol (alcohol)?

Q2 How is ethanol made industrially?

Q3 What is formed when ethanol is burnt in an excess of air?

Q4 What is formed if ethanol is oxidised under controlled conditions?

an alcohol with the formula CH_3CH_2OH

A1 Fermentation using yeast under anaerobic conditions

A2 By reacting ethane with steam in the presence of a catalyst

A3 Carbon dioxide and water only

A4 Ethanoic acid (vinegar), CH_3CO_2H

exam tip You must be able to write the equation for fermentation:

sugar ⟶ ethanol + carbon dioxide

$$C_6H_{12}O_6 \longrightarrow 2CH_3CH_2OH + 2CO_2$$

(47) ANSWERS

GCSE Science & Additional Science: Chemistry
Equilibria and industrial processes

48

Haber process

Q1 Where does the nitrogen used in this process come from?

Q2 Where does the hydrogen used in this process come from?

Q3 What conditions are needed for this reaction to occur?

Q4 What is the yield and how is the ammonia removed from the reacting gases?

ANSWERS

industrial manufacture of ammonia from nitrogen and hydrogen

A1 The air

A2 Either from cracking oil or from the reaction of natural gas (methane) with steam

A3 A temperature of about 450°C, a pressure of 60–150 atmospheres and iron as a catalyst

A4 The yield is about 17%. Ammonia is removed by cooling the gases — ammonia liquefies while unreacted nitrogen and hydrogen remain as gases and are recycled.

***exam* tip** You must remember the equation for this process:
$$N_2 + 3H_2 \rightleftharpoons 2NH_3$$
Carbon dioxide is a by-product of this reaction — it is sold to the drinks industry or used to make urea (a fertiliser).

GCSE Science & Additional Science: Chemistry
Equilibria and industrial processes

Plant oils

Q1 How are vegetable oils extracted?

Q2 Why are vegetable oils important to humans?

Q3 What is an emulsion?

Q4 How is margarine made?

ANSWERS

many plants produce useful oils, vegetable oils being the most important

A1 Fruits, seeds or nuts are crushed and the oil removed by pressing or by distillation.

A2 They provide us with energy and nutrients.

A3 A suspension of one liquid in fine droplets in another liquid with which it is immiscible.

A4 By reacting an unsaturated vegetable oil with hydrogen and a nickel catalyst at 60°C.

exam tip Additives may also be added to plant oils to prolong their shelf-life or improve their appearance. These can be detected by chromatography.

GCSE Science & Additional Science: Chemistry
Equilibria and industrial processes

50

Reversible reactions

Q1 What symbol is used to indicate a reversible reaction is taking place?

Q2 What happens to a reversible reaction in a closed system?

Q3 How do the conditions affect a reversible reaction?

Q4 Do all industrial continuous processes go to completion?

a reaction in which the products of a reaction recombine to form the reactants

A1 The double-headed arrow: ⇌

A2 Equilibrium is reached and the reactions occur at exactly the same rate in both directions.

A3 Conditions affect the relative amounts of all the reacting substances at equilibrium.

A4 No, but enough product may still be made to make the reaction economically viable.

exam tip The yield for an equilibrium is the maximum amount of desired product obtained under the conditions of the equilibrium. For a continuous process, it may not be very high (e.g. ammonia manufacture, 17%).